Food
82

螃蟹大餐

Crabs for Dinner

Gunter Pauli

[比]冈特·鲍利　著

[哥伦]凯瑟琳娜·巴赫　绘

李欢欢　牛玲娟　译

上海远东出版社

目录

Contents

小猪和蘑菇正在讨论为

什么农场池塘里的鱼在逐渐消失。

"一定是有鸟在晚上偷了我们的鱼。"蘑菇这

样认为。

"不，那是不可能的，我们晚上可是很警

惕的，"小猪反驳说，"一定会注意

到的。我想知道在鱼类家族里有

没有吃鱼的种群。"

The pig and the mushroom are
debating why the fish are vanishing
from the pond on their farm.
"It must be the birds who come to
steal our fish at night," argues the
mushroom.
"No, that's not possible; we are very
vigilant," defends the pig. "We would
have noticed. I wonder if one
family of fish isn't eating the
other."

为什么池塘里的鱼在逐渐消失？

Why are the fish vanishing from the pond?

没有鱼对吃其他鱼感兴趣

No fish is interested in eating the other

"看，"蘑菇说，"在3米深的池塘中有7种鱼，每个食物链层次都有一种鱼。没有鱼对吃其他鱼感兴趣，我们可以不用专门为鱼买饲料。那些鱼可能生病了，正在池塘的底部奄奄一息。"

"关于食物链级别，你说得对，顶层的食草鱼类从不吃食底泥鱼类。由于有藻类，水里有许多浮游生物和水蜗牛，这也意味着池塘里所有的鱼都有充足的食物。我实在不明白，过去池塘里鱼的数量是现在的5倍多，发生什么了？"

"Look," answers the mushroom, "we have seven types of fish here – one for each food level of our three metre deep pond. No fish is interested in eating the other. It is possible for us to farm fish without having to buy feed for them. Perhaps they got sick and are dying at the bottom of the pond."

"You're right about the levels. The grass eaters at the top will never bother the bottom feeders. Thanks to the algae, there is a lot of plankton and many tiny snails in the water, which means that there's enough food to feed all the fish. I don't understand it. We used to have five times more fish than we have now, what is going on here?"

"我真的不知道，你觉得是不是池塘里的水出问题了？"蘑菇问道，"我知道你的猪舍里有很多固体物流出来，但这些不是应该被沼气池的过滤器捕获了吗？"

"是的，这些固体物确实是被沼气池捕获了，池塘里的水也来自猪舍。猪粪流入沼气池中，细菌会将水净化，之后才会进入外界。你知道沼气池也生产沼气吗？"

"I really don't know. Do you think there's a problem with the water?" wonders the mushroom. "I know that too many solids flowing out of your sty, but aren't these caught in the digester's filter?"

"Yes, the solids are indeed left behind in the digester. The water for this pond comes from our pigsty. The manure flows into a digester where bacteria purify the water. And the clean water is then returned to nature. Did you know that the digester also creates biogas?"

沼气池也生产沼气

The digester also creates biogas

真有趣！

That's interesting!

"真有趣！"蘑菇说道。

"但如果不是水的问题，那是什么问题呢？"小猪问，"农场里一直都有充足的食物来源。我们也提供了这么多食物，甚至周边曾被认为贫瘠的土壤现在也已经肥沃得超乎人们的期待了。"

"有些人认为贫瘠的土壤会一直贫瘠，这种想法是不正确的。"蘑菇回应道，"有些人认为我们池塘里的水被污染了，因为里面有太多的食物。但池塘流出的水可以神奇地使土壤变得肥沃！"

"That's interesting!" the mushroom replies.

"But if it is not the water, what could the problem be then?" the pig asks. "We have always had plenty of food sources on this farm. We are providing so much food; even the soil around us, which was considered poor for farming, started to flourish beyond everyone's expectations."

"Those who think poor soil will always remain that way, don't think properly," the mushroom replies. "Some people thought the water in our pond was polluted, as there was too much food in it. But the water from the pond turned out to enrich the soil wonderfully!"

"但是，我们还是需要为防止我们的鱼减少而做些什么。"小猪说道。

"所以我们要想一想，池塘里过剩的水会流到哪里？"蘑菇问。

"流到大海里。"

"水流向的海岸带会有什么生物生存呢？"蘑菇想知道。

"But we still need to do something about our fish disappearing, " says the pig.

"So let's think. Where is the excess water from the pond flowing to?" the mushroom asks.

"To the sea."

"And what lives in the coastal zone of the sea where our water flows to?" the mushroom wants to know.

我们要想一想

Let's think

红树林湿地是虾的天堂……

Mangrove swamps are a paradise for shrimp…

"红树林。"

"是的，红树林湿地是虾、螃蟹、海草和海藻的天堂。"

"那你认为其中哪些生物可以从海洋里移到岸上呢？"小猪问道。

"Mangroves."

"Yes, the mangrove swamps are a paradise for shrimp, crabs, seaweed, and algae."

"So which one of these creatures do you think can move up from the sea onto the land?" asks the pig.

"海草和海藻不行，虾也很难。"蘑菇回答道。

"那么，唯一能进入红树林沼泽的就是螃蟹了！"

"Seaweed and algae have no chance, and shrimp can hardly walk," replies the mushroom.

"Well, then the only ones that can walk up into the mangrove swamp, are the crabs!"

唯一能进入红树林沼泽的就是螃蟹了！

The only ones that can walk up are the crabs!

螃蟹吃了我们的鱼？

18

"你的意思是这些走起路来很搞笑的螃蟹吃了我们的鱼？"蘑菇问道。

"我是这样想的。你知道螃蟹是唯一的腿在身体两边而不是在下面的动物吗？"

"You mean those funny walking crabs are feasting on our fish?" the mushroom asks.

"I think so. Do you know that crabs are the only animals who have their legs on the sides of their bodies instead of under their bodies?"

"显然没法阻止他们进入我们的池塘。"蘑菇说道，"他们可能现在就藏在泥土里，随时准备吃掉我们的鱼！"

"让我们把池塘里的水抽干，抓住这些螃蟹吧！当地的农民会非常高兴的，他们可以把螃蟹卖掉赚很多钱，然后供孩子上学。"

……这仅仅是开始！……

"Clearly that has not prevented them from invading our pond," says the mushroom. "They are probably hiding in the mud right now, ready to feast on our fish!"

"Let's dry out the pond and catch all these crabs! It will make our farmer very happy. He will make a lot more money by selling the crabs. And then he will have enough money to pay his kids' school fees."

... AND IT HAS ONLY JUST BEGUN!...

... AND IT HAS ONLY JUST BEGUN! ...

Did You Know ?

你知道吗？

Integrated farming produces fish, agricultural crops, and livestock in such a way that the byproducts of one subsystem becomes a valuable input for another, generating more available food than if each had been cultivated separately.

综合农业生产鱼类、农作物和牲畜，其方法是让某类生物的代谢副产品成为另一种生物的营养源，这比单独培养能够产生更多的食物。

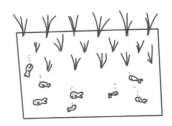

Farming fish in combination with crops and animals means that feed for the fish can be provided naturally and that there is no need to buy fish feed. By doing this, the farmer can save money.

养殖鱼类与种植农作物、饲养动物结合起来，可以自然地为鱼类提供食物而不用专门购买鱼饲料。这样做，农民还可以省钱。

Animals excreta create problems for the environment when not used. Excreta can be used as a source of nutrients and energy for bacteria when they digest the biomass.

动物粪便如果不处理，会带来环境问题。粪便被降解时可以作为细菌的营养物来源。

Fruit and plants could create problems when they rot and become waste. This waste is, however, rich in fibres and an ideal substrate for mushrooms. The waste of mushrooms, the used substrate, is rich in amino acids and therefore a great feed for animals.

水果和植物在腐烂变成废弃物时会产生问题。然而，这些废弃物富含纤维素，对于蘑菇来说是理想的生长基床。蘑菇的废弃物，即它们用过的基床富含氨基酸，是非常好的动物饲料。

There is a lot of food in water and each fish species eats a specific kind of food. Some species like algae, others benthos, and others like grass or diatoms. If there is only one kind of fish in a pond, all the other types of food goes to waste.

水里有许多食物，每种鱼吃特定的食物。有的喜欢海藻，有的爱吃底栖生物，还有的喜欢海草和硅藻。池塘里如果只有一种鱼的话，其他的食物或许就会成为废物。

When several species of fish live in a pond, their excrement creates too much food. The excess biomass can be removed by converting the waste into nutrients for other species. This can be achieved by placing floating flower or rice gardens into the pond.

当不同种类的鱼生活在池塘里时，其粪便会成为更多其他生物的食物。将废弃物转化为其他生物的营养物质，可以去除多余的生物质。这可以通过在池塘里放置水上花卉或种植水稻实现。

Because crabs spend most of their life buried in sand, they have developed flat bodies with long legs on both sides. Some species can walk only sideways because of the way their knees bend.

因为螃蟹大部分时间隐藏在沙子里，所以它们的腿长在扁平的身子两边。一些螃蟹品种膝盖是弯曲的，所以只能横着走路。

Crabs use their pincers to communicate with each other. During mating season, they find comfortable places for the females to release eggs. They work together to provide food and protect their families. Crabs also like to congregate in large numbers to socialize.

螃蟹通过钳子和同伴相互联系。在交配季节，雌性螃蟹会找到舒适的地方产卵。螃蟹喜欢聚集在一起生活，一起工作，为家人提供食物并保护它们。

Think about It 想一想

If fish can find their own food in nature, why do people have to feed farmed fish?

如果鱼在自然界中能够找到食物，为什么人们还要养殖鱼呢？

生活在河流和湖泊里的鱼类通常不止一种。为什么人们只在鱼缸或池塘里养殖一种鱼呢？

There is always more than one species of fish in a river or a lake. Why do fish farmers keep only one type of fish per tank or pond?

If nature cultivates many fish without providing extra feed for them, why do crabs not have the right to feast on it as well?

如果自然界里有很多鱼，但又没有足够的饲料，为什么螃蟹不能捕食鱼呢？

你觉得农民应该是为了生存多挣一些钱，还是也有权利为了娱乐而挣更多的钱呢？

Do you think the farmer should earn enough money only to live, or does he have the right to earn enough to afford entertainment as well?

Let us see who eats what. Start with mushrooms, followed by pigs, and then bacteria, track the nutrition of algae, map the different forms of fish feed, and then figure out who will eat what food in the pond. Can you map the five kingdoms of nature? Once you have figured out all five, try to calculate how much food will eventually be produced if you start with 100 kg of mushroom substrate. This will give you an idea of how productive nature really is.

一起来看看谁在吃些什么。先从蘑菇开始，然后是小猪、细菌，跟踪藻类的营养流动，绘制不同类型鱼饲料的食物链图谱，然后想想它们都吃池塘里的什么食物。你能绘制出大自然的五个生物王国吗？在你找到所有的五类生物后，试着计算一下，以100千克蘑菇生长基床算起，最终能生产出多少食物。这会使你认识到自然界的创造力究竟有多大。

学科知识
Academic Knowledge

生物学	食物链中生物体的营养级别；浮游生物和海藻是初级生产者；鲤鱼和其他鱼种有不同的饮食结构，在鱼塘里一起养殖是互补的；不同种类的螃蟹有不同的饮食结构，所以生活在不同深度的水里；综合生态系统确保了养殖鱼类不需要额外提供饲料——一种生物会为另一种生物提供食物；红树林具有独特的生态系统。
化 学	脂类与聚合物的区别；ω3脂肪酸是含有酯基官能团的多元不饱和脂肪酸，人体自身不能合成；甲烷由产甲烷细菌在厌氧条件下生成；厌氧细菌不能降解木质素；沼气的成分有CH_4、CO_2、N_2、H_2和H_2S。
物 理	综合生态系统利用了重力原理：水从池塘的高处流到低处。
工程学	沼气池的设计可以是单室或双室，沼气池污泥含有高浓度的生化需氧量（BOD）和化学需氧量（COD）。
经济学	养殖户把大部分钱花在资本投资和饲料上，单一养殖鱼的利润很小；在中国，沼气用户量已经有3 000万，沼气池将废弃物转化为清洁能源，增加可支配收入。
伦理学	人们或许认为某些土壤是贫瘠的，但却没想过贫瘠的土壤可以转变成肥沃的土壤。因此，我们限制了自己的思维，常常忽略了我们面前的宝贵资源。
历 史	伏打在1776年发现了厌氧消化；第一个沼气厂1859年在印度孟买建成；1895年生产了第一个通过化粪池沼气发电的灯泡。
地 理	河流和湖泊的沉积物会产生可燃气体。
数 学	线性过程和计算的区别，复杂多变量过程与反馈型非线性计算的区别。
生活方式	许多人不了解食物是怎样生产的，认为土豆来自超市。
社会学	欠发达地区通常以农业为主，进步依赖于就业比重逐渐从农业和工业转变到服务业。
心理学	由于不知道某个东西为何不能运作带来的压力；我们需要分析一个系统的整体环境，以便找出哪个部分不合适，这个系统也被称作"格式塔"（有组织的整体）。
系统论	在食物链中，属于高营养级别的大型食肉鱼类减少后，渔业会使食物网中鱼类减少；厌氧沼气池的设计里含有污水、污泥和粪便的处理过程。

情感智慧
Emotional Intelligence

小猪

小猪具有保护意识，他很伤心鱼的数量在减少，于是开始寻找凶手。小猪明确了答案，通过一步步逻辑分析找出原因。他指出生态系统的优势，并疑惑鱼类在并不缺少食物的前提下数量为什么会减少。蘑菇一连串的问题激发他给出了快速并有针对性的答案，展示了他的洞察力。他对螃蟹身体结构的了解让我们吃惊于他的知识量。小猪认为螃蟹是凶手，对怎样处理这件事提出了可行性建议，这让大家感到很满意。

蘑菇

蘑菇原本想到了一个凶手，但很快就选择了进行详细的分析，找出问题所有可能的原因。他追溯鱼类的食物链，排除了同类相食的可能，开始探讨鱼类减少是不是因为健康问题。蘑菇也提到了水质的问题，他知道一些人会把富营养化的水视为污染物，而这些水对其他生物其实具有高营养价值。蘑菇反驳贫瘠的土壤就一直贫瘠的简单思维。他清晰而果断地表达了自己的观点。蘑菇决定找到答案，问了一系列问题。这帮助他和小猪缩小了可能的凶手范围，得出最终的结论。

艺术
The Arts

你知道鲤科鱼类有多少种吗？不同品种可能看起来很像，但是它们有着自己的食物。为什么不画出来呢：（1）银鲤；（2）鳙鱼；（3）草鱼；（4）鲢鱼；（5）普通鲤鱼；（6）鲫鱼；（7）鲮鱼。然后画一个池塘，以及每一种鱼吃的食物。任何两种鱼的饮食组合都不一样。这些鲤科鱼类的典型食物分别是：（1）浮游植物；（2）浮游动物；（3）植物；（4）底栖生物；（5）昆虫和幼虾；（6）硅藻、蓝藻和植物种子；（7）硅藻、绿藻、植物碎屑和水底腐殖质。你会觉得在池塘里有丰富的食物，每种鱼都会生活得很好。探索各种鲤科鱼类及其食物，会增加你对生物学的理解。

思维拓展
Systems: Making the Connections

综合生态系统按照自然界中的方式模拟营养物质和能量的产生。它展示了大自然中五个生物王国之间的相互联系与不同生物体间营养物质和能量的转移，显示了资源的有效性与生态系统里营养级别的多样性。生态系统越多，就会产生越多的营养物质。一个营养丰富的水体可以被视为一个生态系统。一定区域（如池塘）里多余的营养成分可以供给其他物质（如土壤）。这种生态系统方法与目前饲养业中使用的方法是非常不一样的。目前的方法是，为农业中的每种生物体分别购买饲料，并对每种生物产生的废物分开管理，这增加了经济和环境成本。现代农业注重遗传品质改进和成本控制，综合生态系统保证了所有的营养物质被利用，使现有资源产生更多价值，不需要运输或再处理。这种综合生态系统不仅增加了营养物质的产出，也通过平衡生产和消费，对生物多样性和生态系统健康产生了积极的影响。

动手能力
Capacity to Implement

试想你将经营一个企业，销售罗非鱼鱼种、鱼饲料和养鱼所需的设备。把所有的数字加起来，计算每个人能创造多大价值。以综合农业系统为例，从蘑菇开始，到喂养动物、生产沼气、培育藻类，最终养殖7种鱼，这些鱼能提供大量富含营养的水，使原本贫瘠的土壤变得肥沃。比较一下：不同类型的两种企业哪一家会有更多收入？哪一家会生产更多的食物？

故事灵感来自
This Fable Is Inspired by

陈绍礼教授
Prof. George Chan

陈绍礼在毛里求斯岛出生与长大。第二次世界大战期间他在英国军队服役，之后有机会在英国学习，获得了伦敦帝国理工学院的卫生工程学位。完成学业后，他曾在毛里求斯首都路易港市做城市工程师，后来为驻南太平洋的美国环境保护署工作。1983 年，他来到了祖先的故土中国，并花了 5 年的时间，作为"赤脚工程师"进行综合农业的试验工作。他周游世界，设计了各种各样的农业综合系统方案。他最大的成功在于依靠他的智慧，在巴西把 80 个养猪场转化为能源供应商，并提供肥沃的土地。他参与了 40 多个工程，不仅包括所有的工程设计，还有繁重的体力劳动。他在 80 高龄时仍在参与这些项目，直到后来因为身体原因不得不停止工作，在家乡毛里求斯退休。

图书在版编目（CIP）数据

冈特生态童书.第三辑修订版：全36册：汉英对照／
（比）冈特·鲍利著；（哥伦）凯瑟琳娜·巴赫绘；
何家振等译.—上海：上海远东出版社，2022
书名原文：Gunter's Fables
ISBN 978-7-5476-1850-9

Ⅰ.①冈… Ⅱ.①冈…②凯…③何… Ⅲ.①生态环
境–环境保护–儿童读物—汉、英 Ⅳ.①X171.1-49

中国版本图书馆CIP数据核字（2022）第163904号
著作权合同登记号图字09-2022-0637号

策　　划　张　蓉
责任编辑　程云琦
封面设计　魏　来李　廉

冈特生态童书
螃蟹大餐

[比]冈特·鲍利　著
[哥伦]凯瑟琳娜·巴赫　绘

李欢欢　牛玲娟　译

记得要和身边的小朋友分享环保知识哦！
八喜冰淇淋祝你成为环保小使者！